R03225 82194

T4-AEE-713

```
JUV        Ruiz de Larramendi,
QH           Alberto.
108
.A1        Tropical rain forests
R8513        of Central America.
1993
```

REF

The
Chicago Public Library

CPL

JUV
QH
108
.A1
R8513
1993

Call No. R0095310806

Branch CONRAD SULZER REGIONAL LIBRARY

Form 178

CHICAGO PUBLIC LIBRARY
CONRAD SULZER REGIONAL LIBRARY
4455 LINCOLN AVE.
CHICAGO, ILLINOIS 60625

JAN 1994

BAKER & TAYLOR BOOKS

THE WORLD HERITAGE

TROPICAL RAIN FORESTS OF CENTRAL AMERICA

UNESCO

CHILDRENS PRESS®
CHICAGO

Table of Contents

Introduction 4
A Bridge between the Americas 6
Darién National Park 12
On the Border between Two Worlds 12
The Riches of the Forest 14
Some Rain Forest Terms 14
The Cuna and Choco Indians 16
On the Trail of El Dorado 18
The Talamanca-La Amistad Cordillera 22
A View of Two Oceans 24
Tropical Timeline 24
Some Neotropical Mammals 26
The Río Plátano Biosphere Reserve 28
The Mangrove Barrier 28
The Destruction of the Tropical Rain Forests 30
Glossary 32
Index 33

Library of Congress Cataloging-in-Publication Data
Ruiz de Larramendi, Alberto.
 [Bosques tropicales de America Central. English]
 Tropical Rain Forests of Central America / by Alberto Ruiz de Larramendi.
  p. cm.—(The World heritage)
 Includes index.
 Summary: Describes several rain forests in Central America and discusses the plants, animals, and Indians that live, or have lived, there.
 ISBN 0-516-08383-X
 1. Rain forests—Central America—Juvenile literature. 2. Rain forest ecology—Central America—Juvenile literature. 3. Natural history—Central America—Juvenile literature. 4. Indians of Central America—Juvenile literature. [1. Natural history—Central America. 2. Rain forests—Central America. 3. Rain forest ecology. 4. Ecology. 5. Indians of Central America.] I. Title. II. Series.
QH108.A1R8513 1993
574.5'2642'0972—dc20   92-35062
                                                                      CIP
                                                                      AC

Bosques tropicales de America Central: © INCAFO S.A./Ediciones S.M./UNESCO 1991
Tropical Rain Forests of Central America: © Childrens Press, Inc./UNESCO 1993

ISBN (UNESCO) 92-3-102682-1
ISBN (Childrens Press) 0-516-08383-X

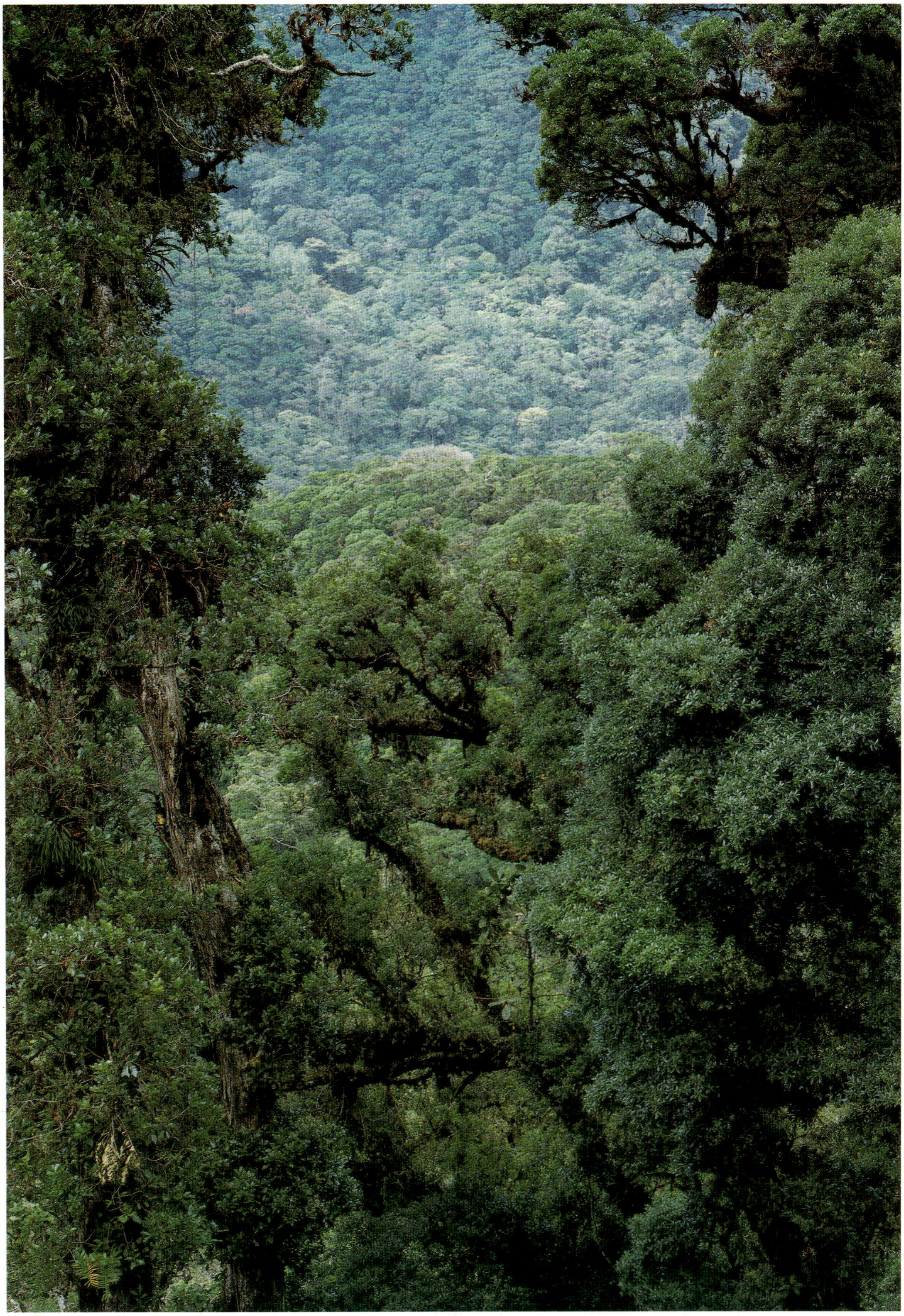

Tropical Rain Forests of Central America

Central America is a land of adventure. The accounts of the early explorers blend with the half-hidden traces of cultures that existed long before Columbus reached the New World. The well-known tales of Cortés, Pizarro, Cabeza de Vaca, and others contrast with the ruins of ancient Indians. Even the customs of the Indians who survive today—the Cuna, Miskito, and Choco—are still engulfed in mystery.

But Central America, and especially the World Heritage sites presented here, also has impressive natural wonders to entice the curious traveler. Remarkable animal life and incomparable forests make this region a paradise—one of the last of its kind on earth.

Threatened with Extinction
Tropical rain forests are among the most severely threatened ecosystems on earth. Their systematic destruction throughout the American continent alarms conservationists throughout the world. These photos show two protected rain forests of Central America: Darién National Park in Panama *(left)*; La Amistad National Park in Costa Rica *(right)*.

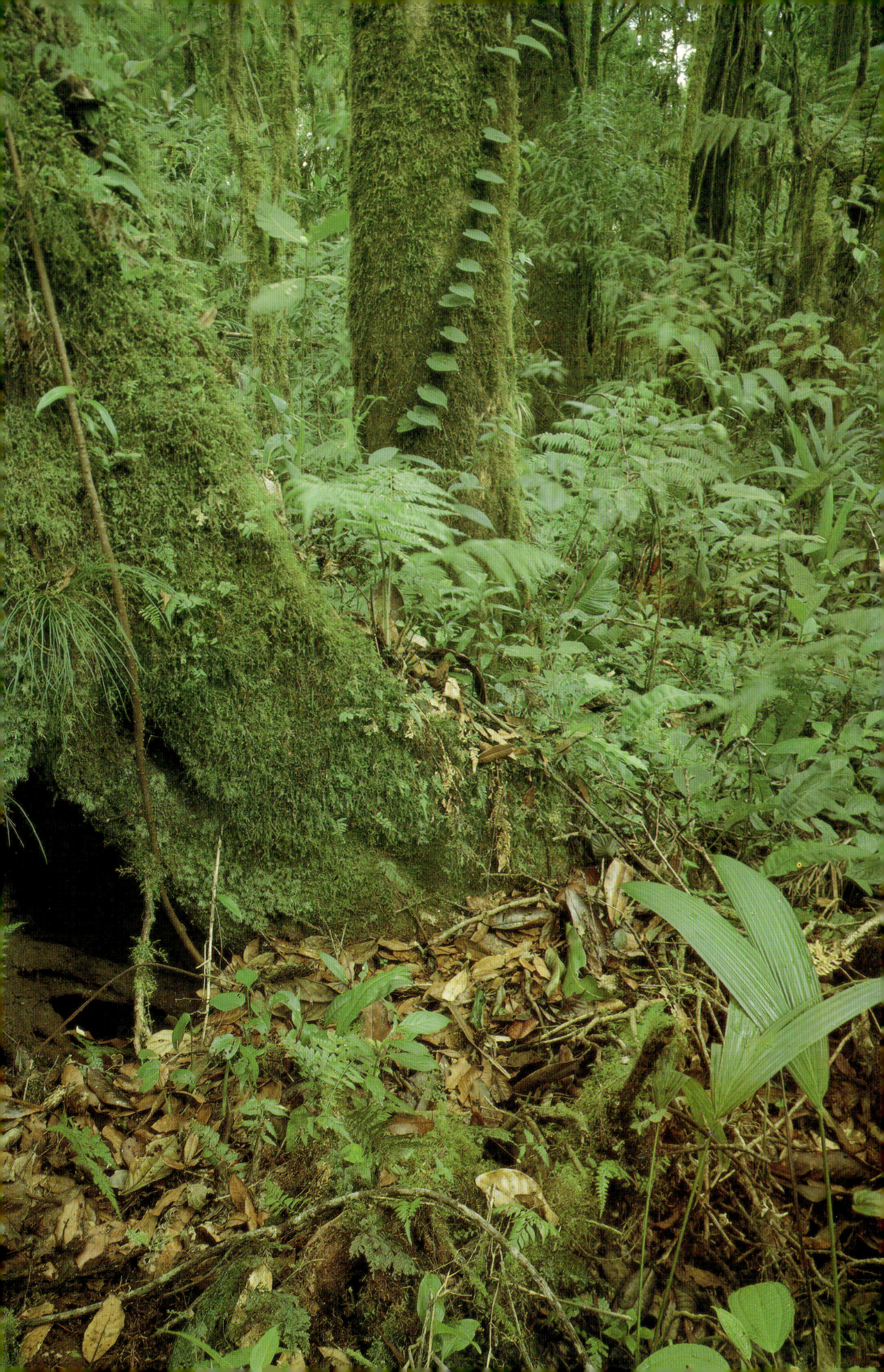

A Bridge between the Americas

A quick glance at a map shows that the three World Heritage sites highlighted in this book are very close to one another. (See map on page 10.) They are all located in Central America—the narrow neck of land, or isthmus, that separates North America from South America. Here live plants and animals native to both continents.

It might seem that living things moved freely back and forth along the isthmus throughout the ages. Actually, however, most of the migrations have been in one direction, from north to south. The two continents have not always been connected, either. Several times during the earth's history, Central America has been submerged beneath the ocean, only to rise again. To understand the plants and animals now living in Central America, we must study the long history of this corridor between the continents.

According to the theory of plate tectonics, the earth's surface is made up of a series of rigid plates. These plates have been shifting around throughout the earth's history. Scientists used to call this "continental drift." But now it seems that much more has been moving around than just the continents. The theory of plate tectonics argues that these plates rest on soft, viscous material that surrounds the earth's core. Over millions of years, some of the plates rose, forming continents, and others sank beneath the seas. The plates do not fit together precisely. There are many cracks between them, and some of the viscous material manages to squeeze out. This material cools and hardens, forming a thick ridge on the edges of nearby plates. One such ridge exists in the middle of the Atlantic Ocean. It helped to bring about the gradual separation between North America and Europe.

At other points where plates collide, the edge of one is pushed beneath the rim of another. At times the pressure of this movement creates mountain ranges. A line of contact between plates runs along the Pacific coast of North and South America, forming the Rocky Mountains and the Andes.

Río Plátano Biosphere Reserve

One of the last jungles of Central America is located in Honduras, in a forest preserve in the Plátano River basin *(opposite page, bottom photo)*. This area has been designated a biosphere reserve by UNESCO. About two thousand Miskito Indians live here, mostly in small villages *(upper right)*.

The Choco Indians

Central America's rain forests not only preserve a rich and varied plant and animal life. They also shelter native peoples perfectly adapted to these surroundings. The Choco Indians are among the people who live in Darién National Park *(left)*.

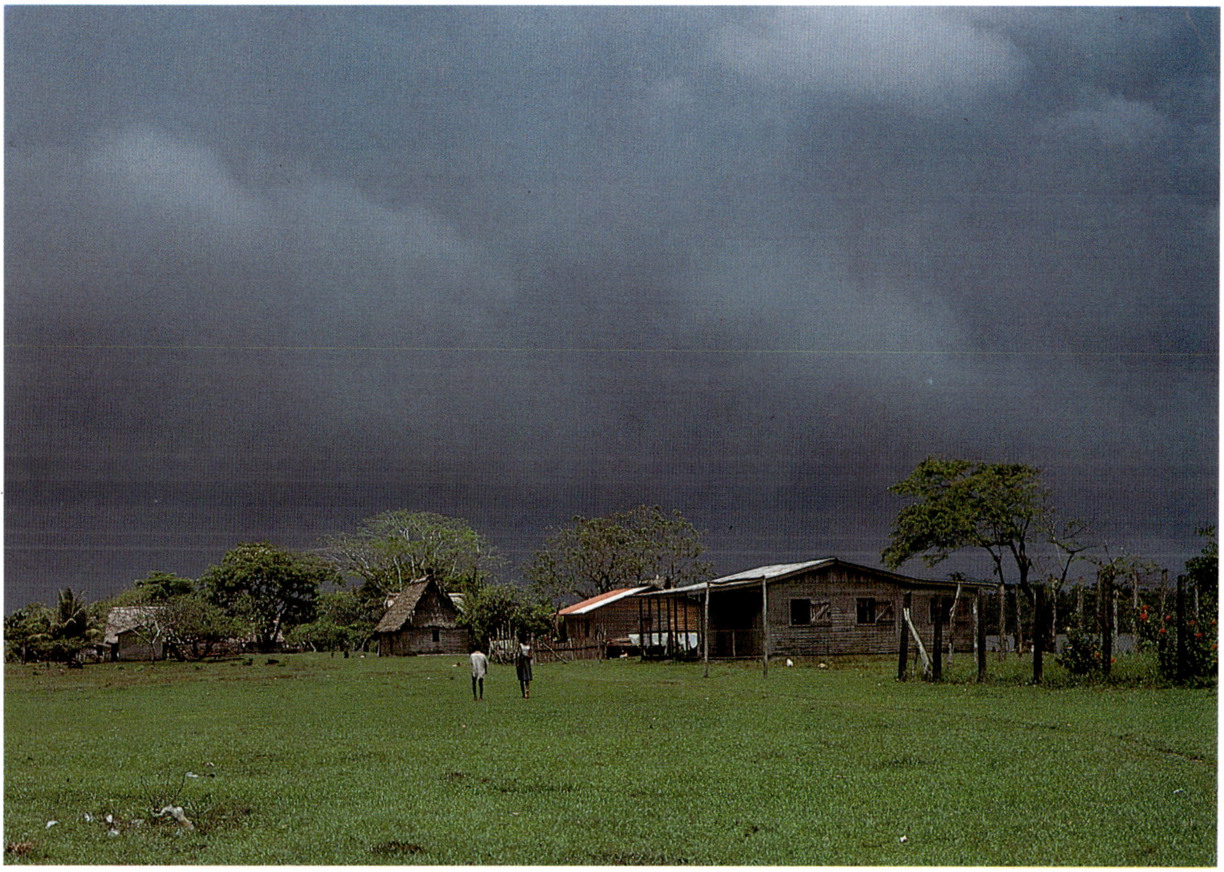

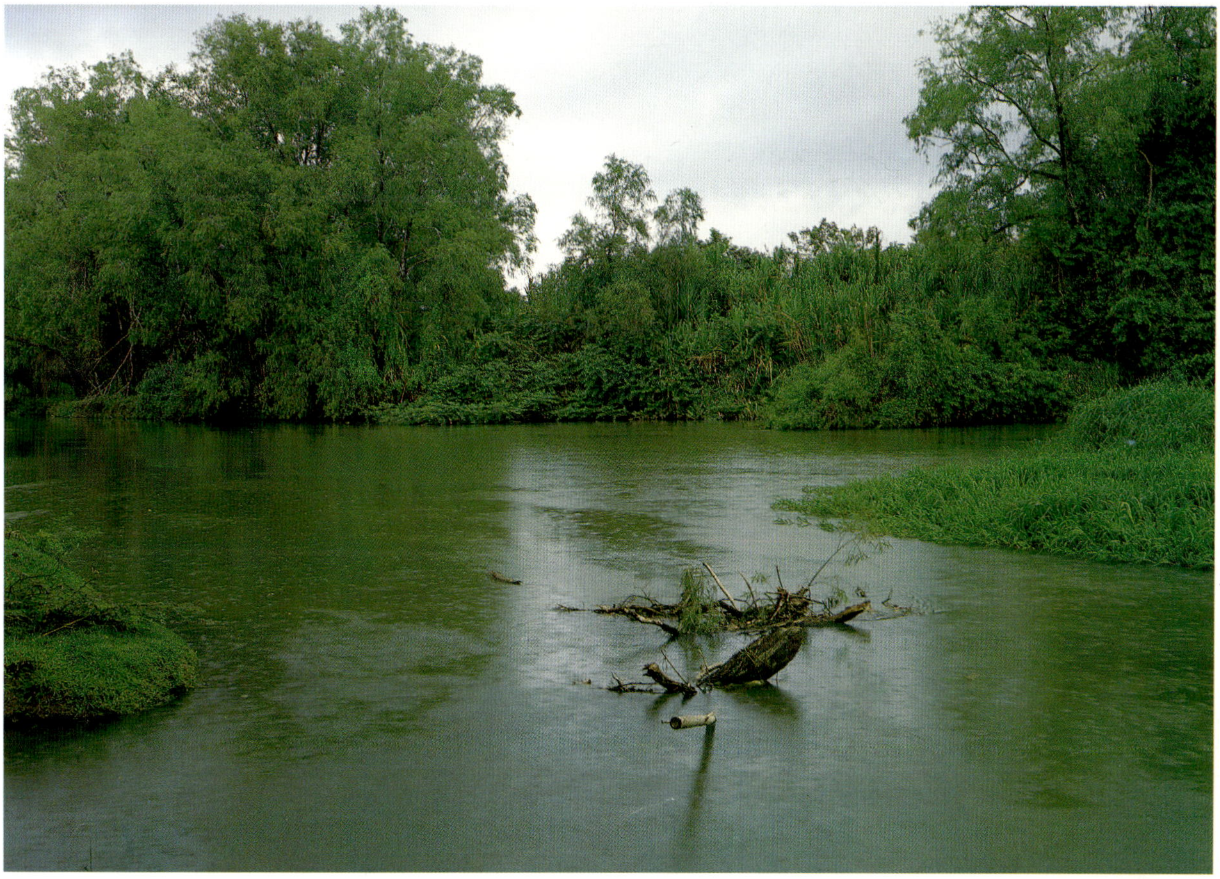

This theory—still debated by scientists—also holds that South America was once joined to Africa. The continents were arranged in this way at least until the end of the Jurassic Period, about 135 million years ago. Then South America began its slow journey to its present location. It reached that point about 65 million years ago, near the end of the Cretaceous Period. During this time, the two Americas were divided by a narrow arm of the sea. Gradually this gap was filled in with silt.

By about 60 million years ago, the narrowest part of Central America—where Costa Rica and Panama are now—formed a land bridge between the two American continents. This tongue of land allowed a host of animals from North America to migrate all the way through the Central American jungle into South America. Many South American animals migrated northward, too. Following this invasion, Central America was partially submerged under water once again, forming a chain of islands. This put a stop to the migration of most land animals, especially the large northern mammals.

Now the animals of Central and South America had the chance to develop independently, cut off from the rest of the world. A line of marsupial mammals evolved into carnivores, or meat-eaters. One of these was very similar to the sabertoothed tiger. Various placental mammals developed into such amazing creatures as the *Toxodon* (much like a hippopotamus), the *Pyroterium* (resembling an elephant), and the horse-like *Thoaterium*. One marsupial surviving to this day is the opossum, a legacy of those distant times. Other survivors include insect-eating placental mammals such as the armadillo, the sloth, and the anteater.

The existence of a chain of islands allowed some species to reach as far as South America. This group included some monkeys and rodents. In South America a species of rodent, the capybara, evolved into the largest-sized rodent on earth. Capybaras can weigh as much as 100 pounds (45 kilograms)!

About three million years ago, the Panamanian isthmus solidified into the form we know today. Now it served as a path between the Americas. An avalanche of northern species— specialized by the hard, competitive process of evolution— reached the southern subcontinent. These included the fox, coyote, jaguar, mountain lion, and other cats. Various new arrivals, members of the dog and cat families, wiped out the old marsupial predators. Opossums and mouse-possums remained, while deer-like and camel-like animals spread over the lowlands and mountains of South America. Tapirs, peccaries, and new types of rodents replaced earlier creatures.

A Wealth of Animals
Besides the rich plant life of the Central American rain forests, there is also an enormous abundance and variety of animals. Hundreds of species of invertebrates (creatures without spines) have not even been described yet. The vertebrates, such as birds and mammals, are also very well represented. These photographs show two typical species of Central America's tropical jungles: the tranquil sloth and the beautiful toucan with its sulphur-colored throat.

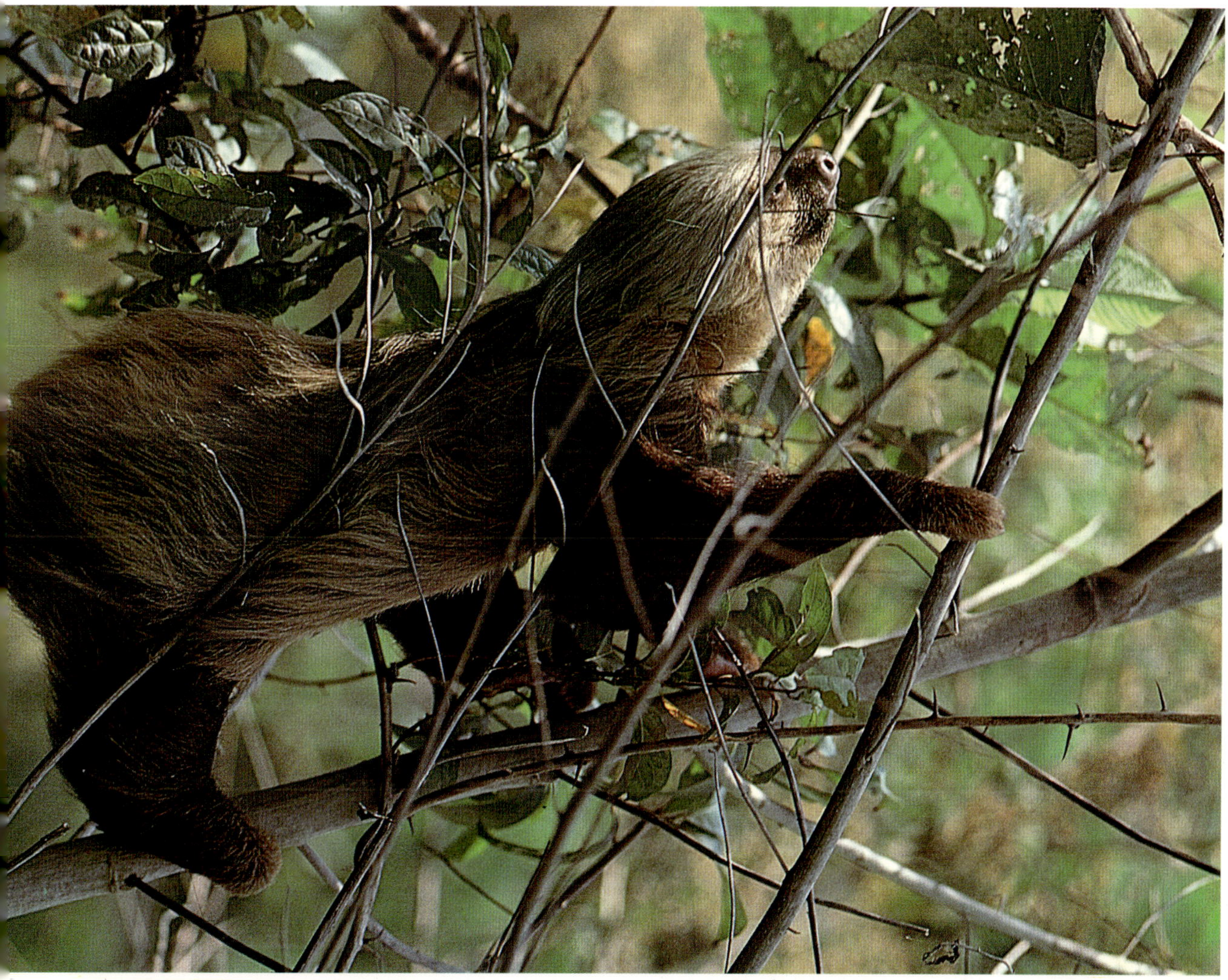

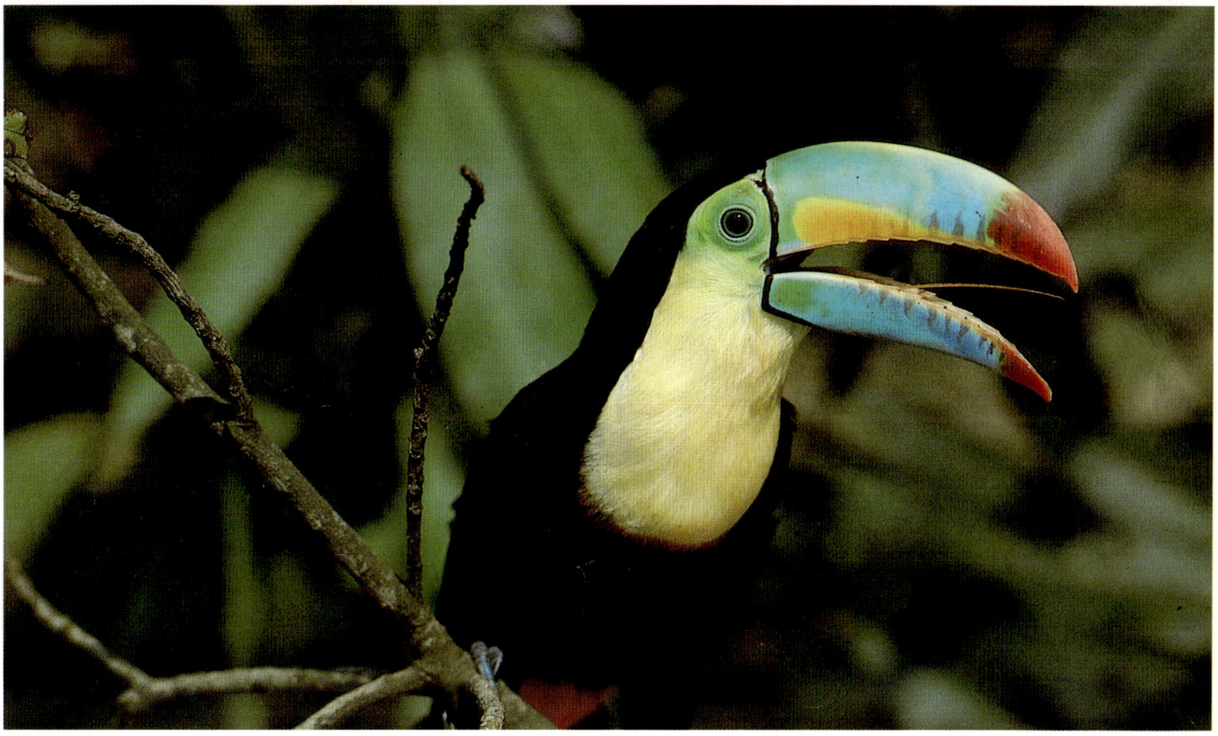

Central America has witnessed the comings and goings of many animals between North and South America. This is especially true in the three World Heritage sites featured in this book. In all, about fifteen families of mammals crossed from North to South America. Meanwhile, only about half that number moved the other way.

Animals that moved into Central and South America found a generous climate and plenty of food. Over time, the original animals developed new variations. As a result, Central and South America have some of the richest animal life on earth.

Although vast numbers of species became extinct, some animals from the early periods still exist. Modern animals now live side by side with "living fossils"—animals that have stayed the same through many periods in the earth's history. Studying them helps us to learn about some of the principal sagas in the plant and animal life of the planet.

In Central America there are numerous forms of life typical of North America. But the animals of Central America are most similar to those of South America. Both Central and South America lie in a biological and geological region called the Neotropical zone. Included in the Neotropics are South America, the West Indies, and tropical parts of North America.

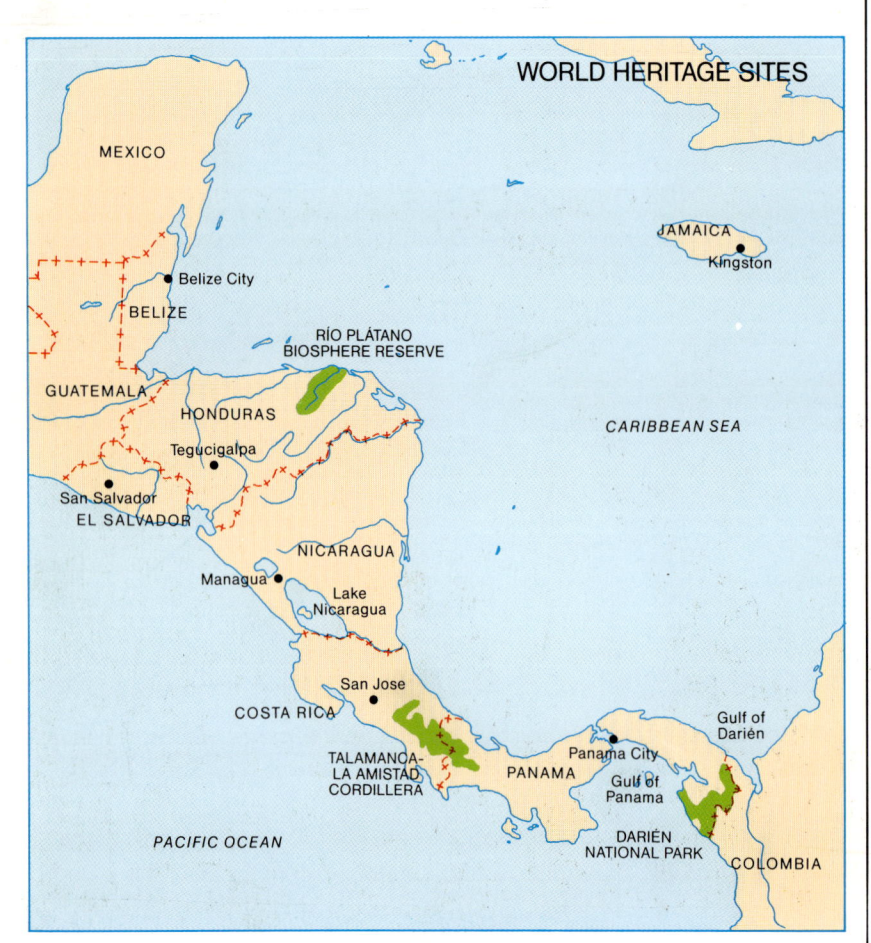

Gold Fever
Before he came upon the Pacific Ocean, Vasco Núñez de Balboa crossed the jungles of Darién National Park. Later many men, searching for El Dorado, tried to find gold in the soil of these dense tropical forests. This old abandoned mine in the Cana region (*opposite page, top*) is a reminder of the gold fever that once swept this Panamanian national park. Below it are two typical plant species in the same protected area.

11

Darién National Park

September 1, 1513. A small army commanded by Vasco Núñez de Balboa departs from the little settlement of Santa María la Antigua del Darién. Balboa had been warned that there were serious charges against him back in Spain. If only he could make some great discovery—gold or perhaps even the legendary South Sea—he might be able to win the king's favor.

Balboa's men sail along the northern coast of present-day Panama for a few days, then begin marching inland. After an exhausting trek through jungles and swamps and a fierce battle with local warriors, an Indian guide leads Balboa up a peak. Here, at last, stretched that endless ocean, the great South Sea. (It was not until six years later that another explorer, Ferdinand Magellan, named it the Pacific.) Balboa immediately recognized the importance of this discovery. Columbus's original idea of sailing west to reach the Indies was indeed correct. But he had bumped into an unforeseen difficulty: an entire continent—America—that lay between the two known worlds. Now, finally, Balboa gazed at the great ocean upon which many future explorers would sail to the Indies.

On the Border between Two Worlds

Darién National Park, created in 1980 upon 1.4 million acres (570,000 hectares) of land, straddles Panama and Colombia at the southern end of the Central American isthmus. Almost the entire park lies in the mountains that slope down toward the Pacific coast. The waters of the Bayano, Chucunaque, and Tuira rivers pour down its mountainsides to the ocean. The region contains one of the least disturbed tropical rain forests in Central America.

Darién National Park protects some exceptionally valuable natural resources, and it also serves as a protective barrier. All along the western edge of North and South America runs a route called the Pan American Highway. The Inter-American Highway, the section that runs through Central America, is interrupted in Darién. This is called the Darién Gap. The absence of roads through Darién prevents South American cattle diseases from traveling northward. Finally, the park is a natural barrier against the drug trade and illegal traffic of endangered species (chiefly parrots, orchids, and the skins of reptiles and spotted cats). Originating in several countries of Central and South America, this illegal trade relies in part on some of the capital cities of Central America for its worldwide exports.

The Jungle Environment
The jungles of Darién are considered among the most diverse in tropical America. Many rivers crisscross the national park, including the Perensemico River (*opposite page, top*). Two totally different indigenous native cultures exist in Darién: the Choco and the Cuna. Both these peoples are adapted to life in the unique environment where they have settled. (*Bottom*) A Cuna home.

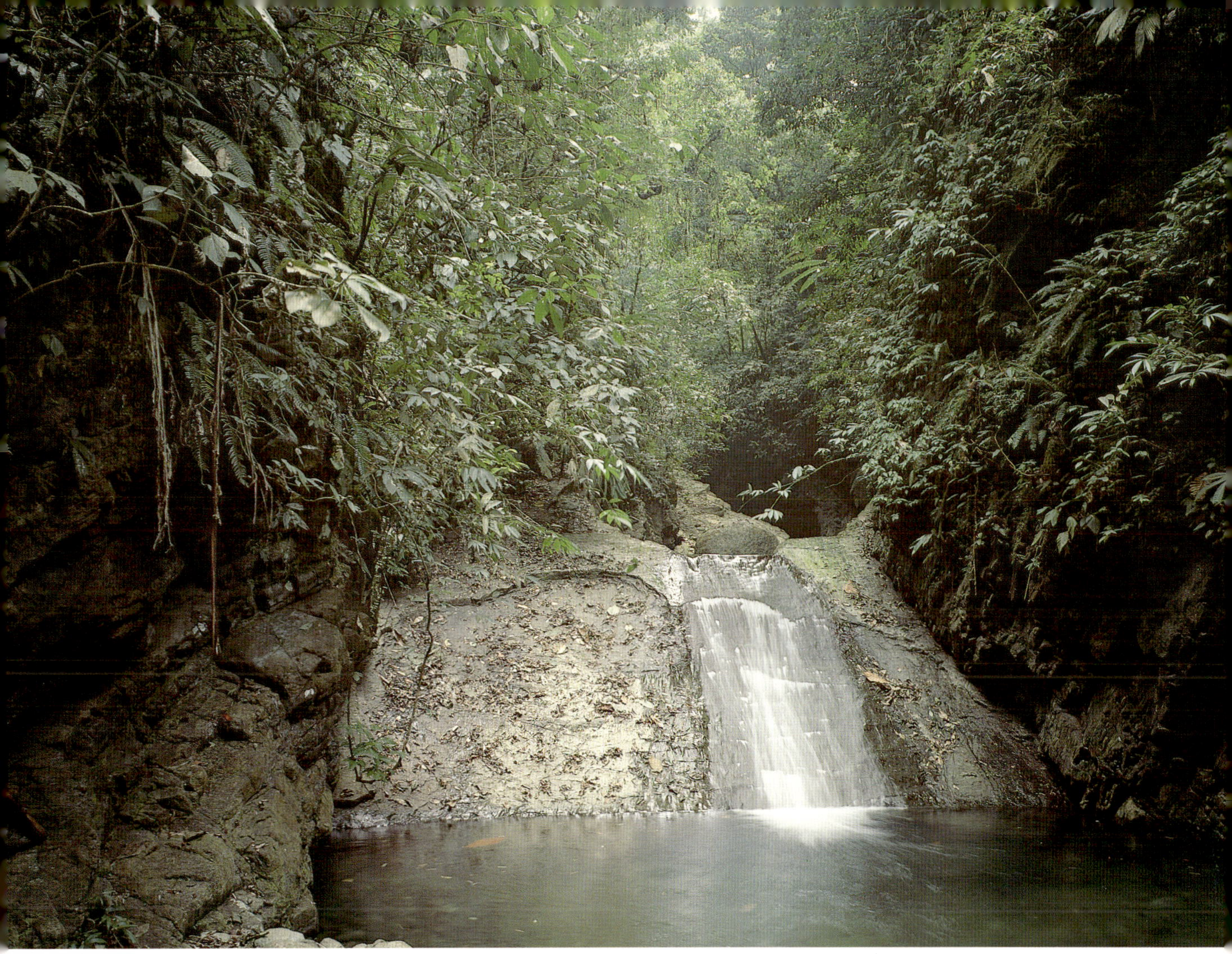

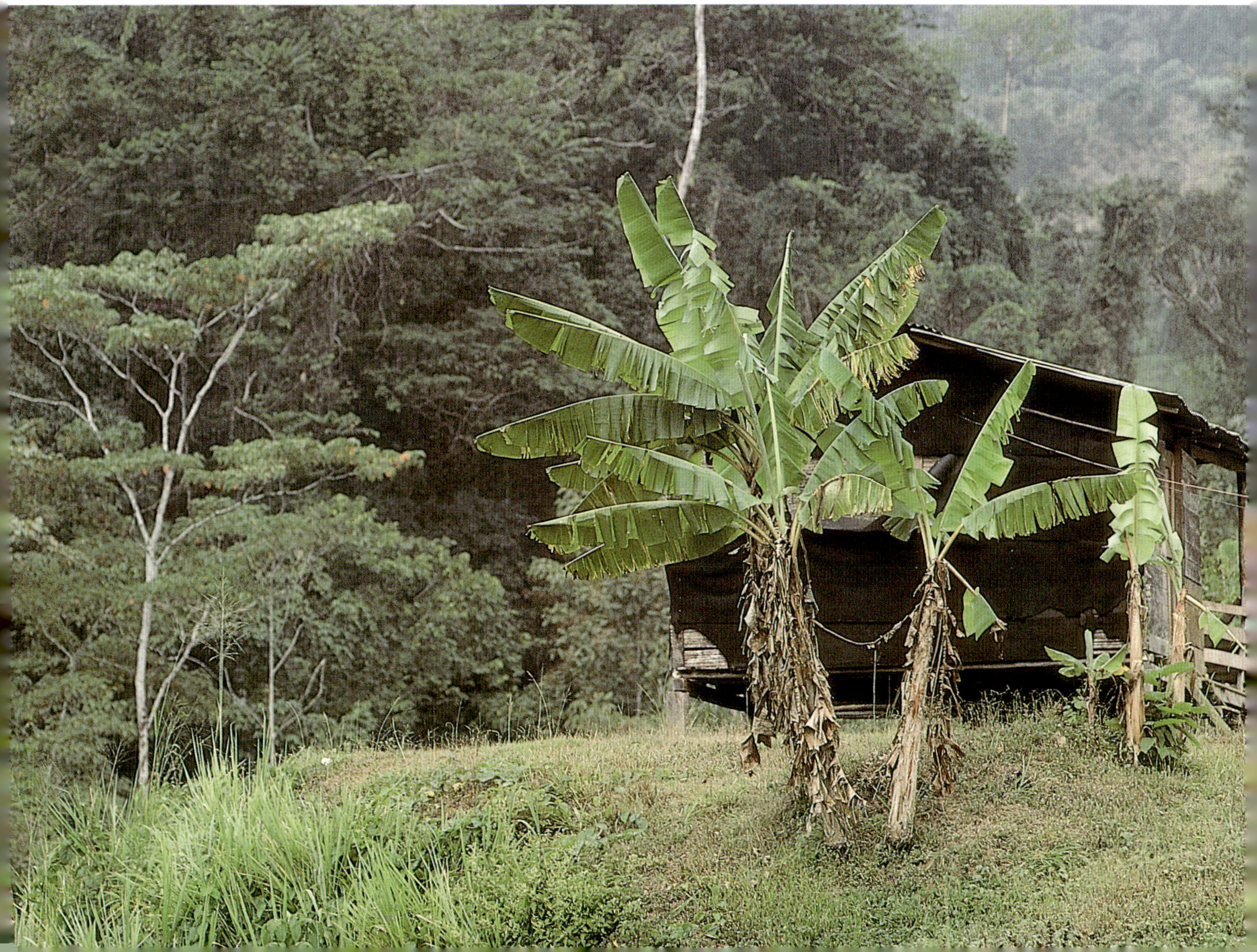

With these problems in mind, Darién park has been divided into three distinct sections. The zone of absolute conservation embraces all of the areas above 1,300 feet (400 meters). The cultural zone is preserved for the indigenous people. In the zone of controlled development, a limited amount of travel and ecological change is allowed.

The Riches of the Forest

Any gardening enthusiast knows that plants grow best when three factors come together: good soil, plenty of water, and a warm, steamy climate.

All of these conditions are found in Darién. The soil is amazingly fertile. It is made up of limestone and sandstone, with bits of volcanic rock and ash. The climate is warm and humid. About 120 inches (300 centimeters) of precipitation falls in Darién every year. Even when it isn't raining, the air in Darién is very moist. Thick mists drifting in from the ocean sometimes hang over the mountains. The temperature hardly changes year round.

The park's plant life changes at different altitudes. Mangrove swamps flourish along the coast, while the cloudy mountain peaks are covered with almost unexplored dwarf forest. (With less oxygen in the air at high altitudes, trees do not grow as tall.) In the wet lowlands are marshy forests of *cativo*, prized for its wood, while in more arid areas the *cuipo* is the most common tree.

Some Rain Forest Terms

biosphere: a system that consists of living creatures plus their environment

conservation: preserving and protecting an area to keep it from being destroyed

degenerate: to decline and fall apart

ecology: the relationship between living things in a certain area and their environment

ecosystem: a system of living things and their environment working together as a unit in nature

habitat: the kind of place where a certain animal or plant usually lives

indigenous: native, natural, or original to a region

Neotropical region: one of six major regions of the earth defined by its typical plant and animal life. The Neotropics includes South America, most of Central America, the West Indies, and tropical parts of North America. Typical Neotropical animals are llamas, tapirs, jaguars, opossums, and a rich assortment of insects and birds.

Ecological Niches for All
Trees in the Central American rain forests may grow to be more than 130 feet (40 meters) in height. Thus the forests provide a wide range of habitats, from the leafy canopy to the underbrush. These habitats, or ecological niches, permit many species to live close together. (*Right*) One of the most beautiful birds of the Central American jungle, the yellow-and-blue macaw.

15

In addition to monkeys, rodents, tapirs, and reptiles, Darién is home to many spectacular birds. Two of its carrion birds are found throughout the Americas, the turkey vulture and the black vulture. Two others are found only in Neotropical areas—the king vulture and the yellow-headed vulture. Among the birds of prey are the osprey, the solitary eagle, and the harpy eagle. The harpy eagle is the central figure on Panama's coat of arms, a tribute to this majestic creature. It is the largest and most powerful of all eagles on the planet, with a wingspan that can reach over 6 feet (2 meters).

The Cuna and Choco Indians

Animals are not the only ones to migrate through the Central American isthmus. Anthropologists estimate that, about 10,000 years ago, the first Amerindians—descendants of people who migrated from faraway Asia—reached Darién. Today the Indian population of the Darién area numbers around 5,000 people, members of the Cuna and Choco groups.

Most of Panama's Cuna live in the *comarca* (reservation) of San Blas, comprising the islands and coastal lands along the Caribbean. But a few thousand live in small villages on the banks of the Chucunaque, Bayano, and Tuira rivers in Darién.

Cuna society is matrilineal—that is, the women own the land, and daughters inherit the family property passed down on the mother's side. When a young couple marries, they live with the wife's family.

Cuna men work at farming, hunting, and fishing. Using slash-and-burn farming methods, they raise coffee, cocoa, corn, and ivory nuts. They also sell coconuts in the markets of big cities. To fish, they use hooks, nets, sharp sticks, and fishing corrals or wooden traps called *empalizados*. They hunt wild pigs, iguanas, and wild turkeys.

Selling *molas* is an important source of income for the Cuna. *Molas* are square cloth panels with colorfully embroidered and appliqued patterns. Cuna women wear *molas* as blouses, and the panels are in great demand by tourists and other shoppers.

Important figures in Cuna society are the *nele*, or spiritual leader; the *innatuledi*, or herbal healer; and the *absoguedi* or witch doctor, who forewarns of catastrophes. The *kantule* is the honored wise man, historian, and priest of a Cuna group.

Children of Darién
The Choco Indians (*opposite page*) first entered Darién at the time of the Spanish conquest and have lived there ever since. They hardly number one thousand people, scattered in small settlements along the riverbanks. They grow much of their own food and also gather wild fruit. Their well-ventilated houses are built on wooden stilts and have cone-shaped roofs. The women are very skilled in the art of basketry. Choco artisans do elaborate work in wicker, wood, and textiles.

17

Choco Indians migrated into the Darién region from northern Colombia, where most Choco still live. They reached Darién during the Spanish conquest and managed to fight off the Spaniards several times. Less influenced than the Cuna by outsiders, they live in villages along the riverbanks. They navigate the area's many waterways in *piraguas*, or dugout canoes. To catch fish, they cast harpoons from their *piraguas* or use machetes while swimming underwater. They hunt with bow and arrow or with blowguns that shoot poisoned darts. Around their homes the Choco grow medicinal plants and various food crops such as rice, maize, and sugarcane. They grow plantains to sell. The Choco have a patrilineal society, in which property is passed on through the males.

Roads through the Jungle
The denseness and thin soil of the Central American rain forests make it difficult, and at times impossible, to travel in motor vehicles. The ample rivers of the Río Plátano Reserve (*opposite page, top*) and Darién National Park (*bottom*) are the real highways of the jungle. Thanks to these waterways, native residents use rustic canoes to reach the villages along the banks.

On the Trail of El Dorado

Costa Rica ("rich coast") owes its name to the Spanish conquistadors. When Christopher Columbus arrived there in 1502, he was dazzled by the splendid gold ornaments the natives wore. The Darién region, and Cana in particular, yielded some of the most important gold deposits in Latin America, after the Aztecs' in Mexico and the Incas' of Peru. The Cana mines were located near the source of the Tuira River, at the Panama-Colombia border. The Indians developed a culture based on gold, which they crafted exquisitely. The region supplied the Indians with gold until the Spaniards took control of the region and its mines.

The legend of El Dorado (the Golden Man) was based on a custom of the Chibcha Indians, who lived high in the Andes Mountains of Colombia. At the coronation of a Chibcha prince, his body was completely covered with gold dust. Then he was rowed out into a lake, where he plunged in, making the gold dust an offering to the spirits. The Indians also threw statues of solid gold and precious stones into the water. On this point, legend and reality blend together. In Lake Guatavita, in the Colombian Andes, a number of these objects have actually been found. In fact, these artifacts tell us much of what we know about Chibcha culture.

Tales of El Dorado lured many adventurers to explore deeply into the wilds of the New World. "El Dorado" came to mean a land whose very streets were paved with gold. Spaniards first heard of the legend before 1530. For more than two centuries afterward, expeditions set out in search of the fabulous treasure that awaited them there. In 1537, Gonzalo Jimenez de Quesada was sent to find El Dorado. In 1538 he conquered the Chibcha and founded the city of Bogota, now the capital of Colombia. Gonzalo Pizarro (1539), Francisco de Orellana (1541), England's Sir Walter Raleigh (1595), and several German explorers were among those who risked their lives for gold fever. It is said that a large portion of Central and South America has never again been explored as thoroughly as it was at that time. Nevertheless, the adventure ended with more deaths than riches. As one author put it, "The hungry throat of El Dorado had an insatiable appetite for the souls of heroes."

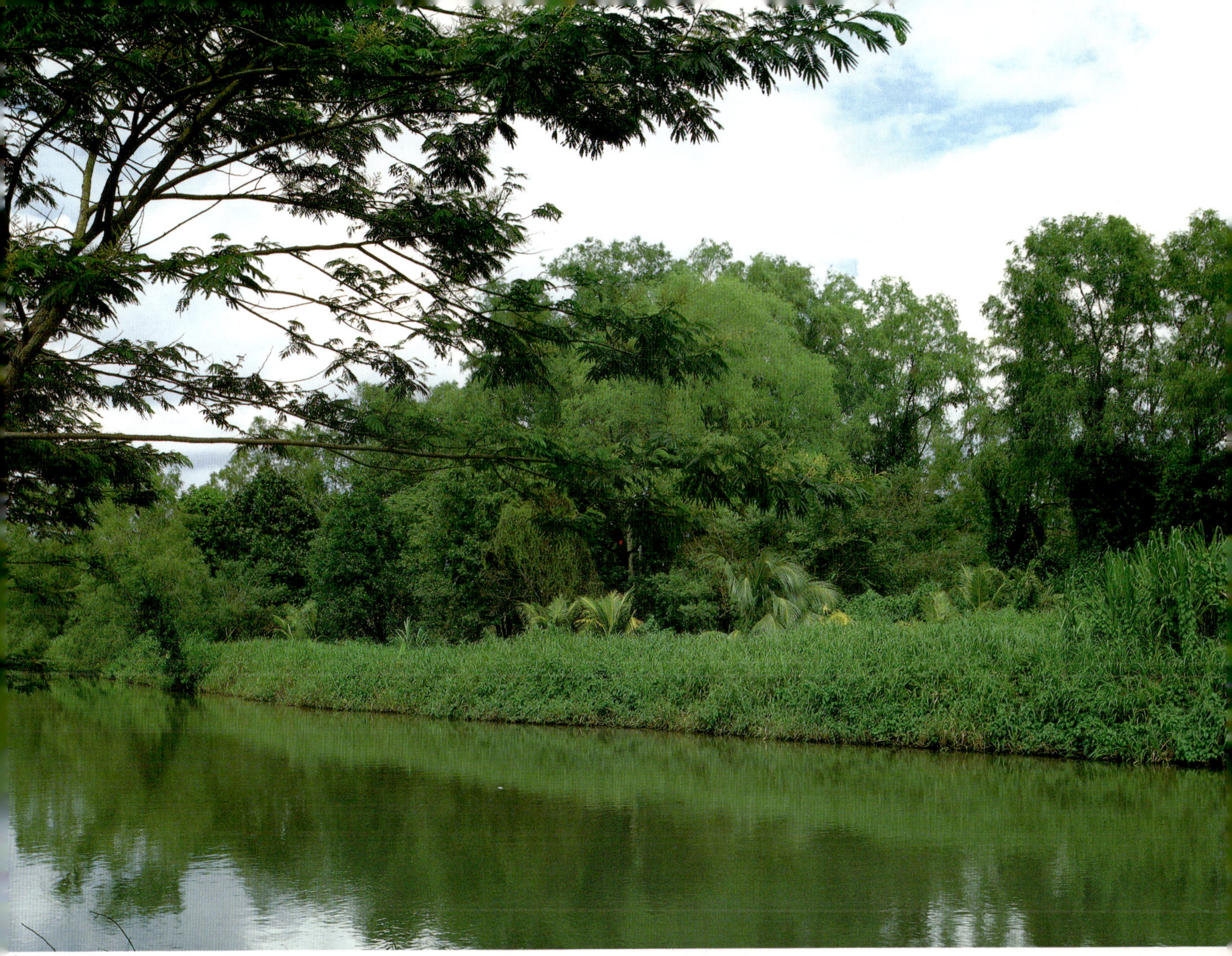

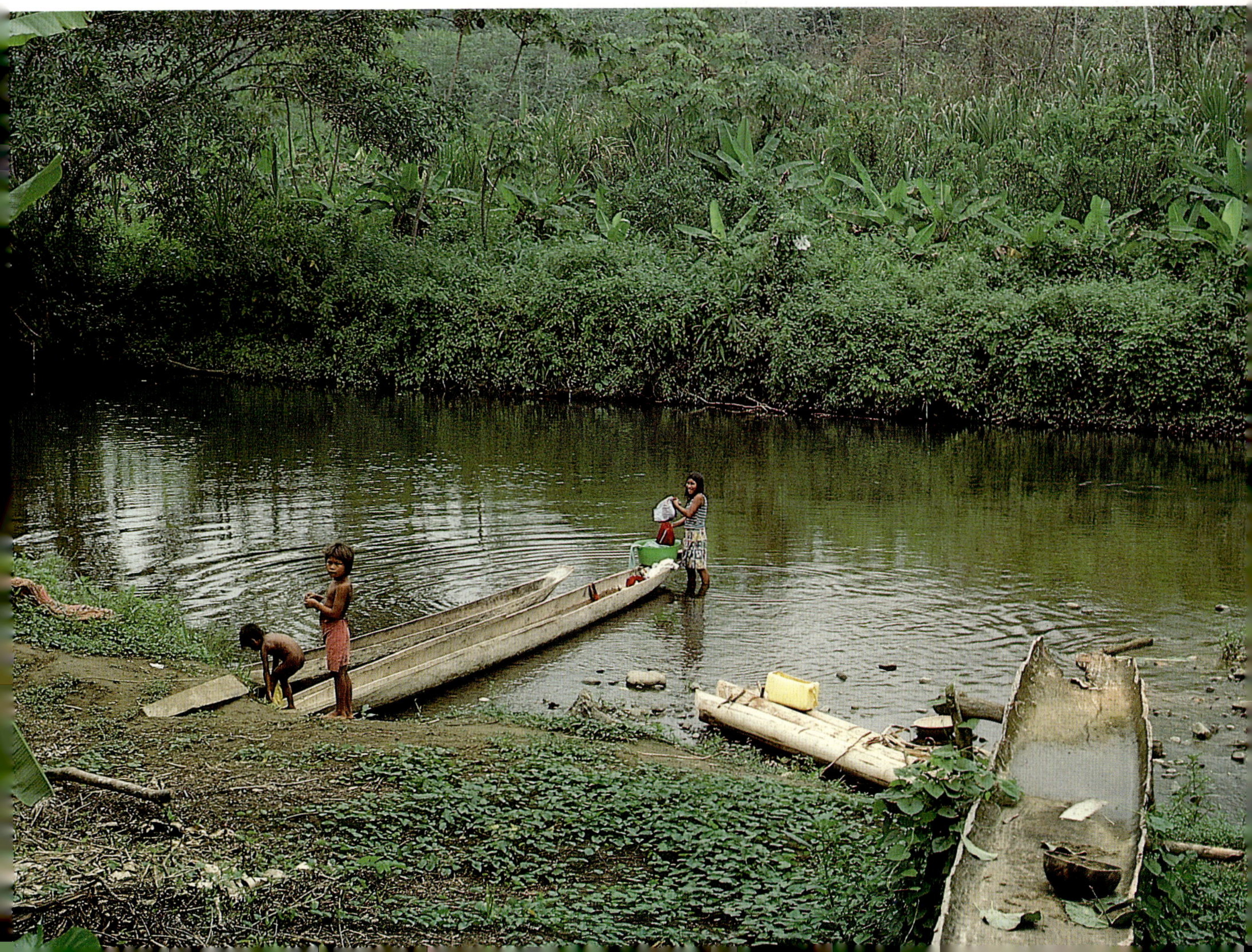

A Wildlife Refuge
Many animals in danger of extinction find their last refuge in the forests of Central America. These forests are like islands surrounded by territory that has been transformed by industry and by destructive farming methods. Species such as the puma (*right*), which has disappeared from many parts of Central America, have stable populations today in Darién National Park and in the Talamanca-La Amistad Cordillera.

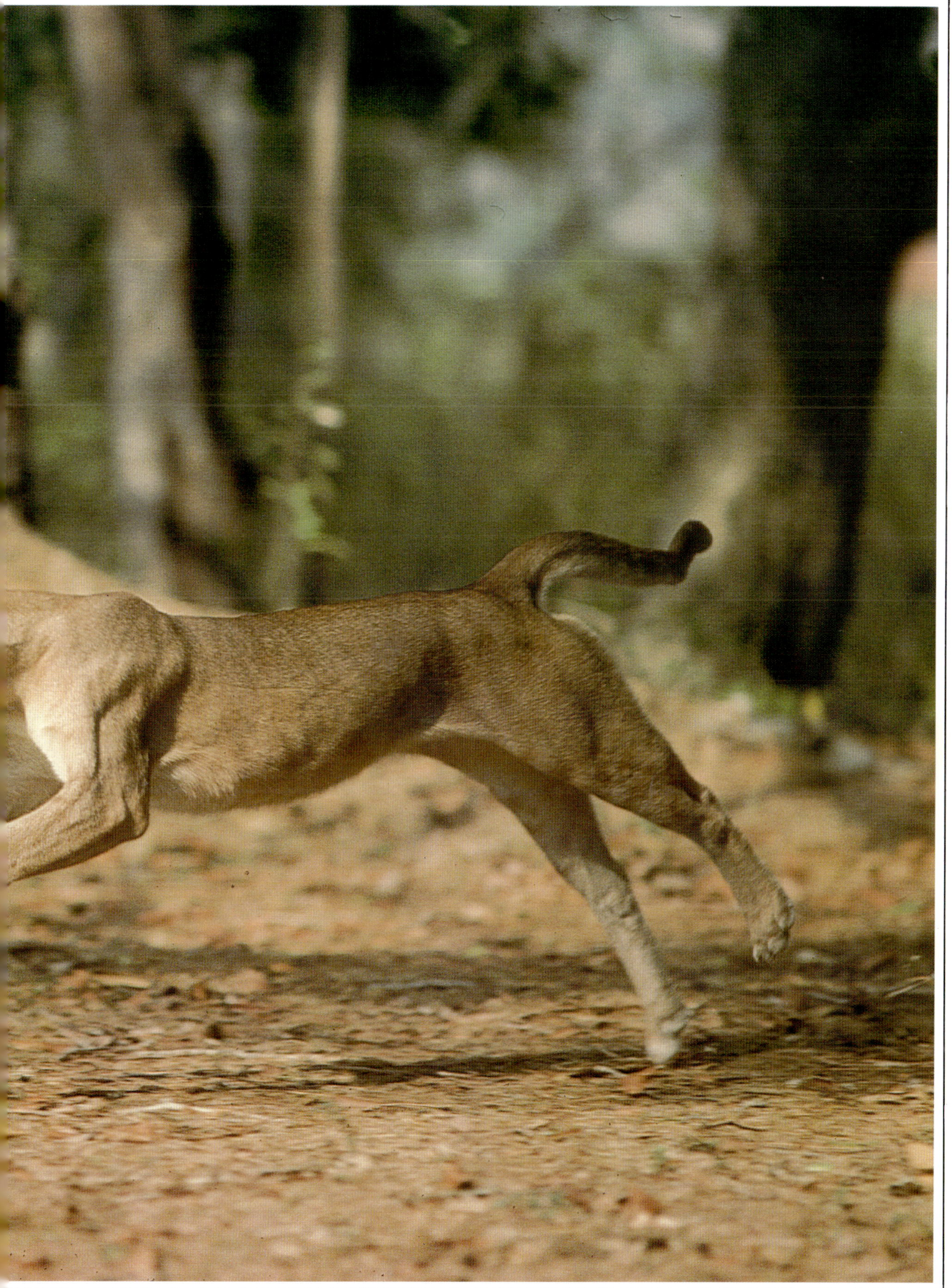

Anthropologists lament the tragic history of indigenous American peoples such as the Cuna and the Choco. Many groups have been forced to trade their ancient, richly spiritual values for Western culture. Perhaps this is unavoidable. But it is encouraging to know that in certain protected areas, such as Darién National Park, these communities can maintain a vital link with their ancestral past.

The Talamanca-La Amistad Cordillera

Rib-like mountain ranges run through most of the Central American isthmus. From their slopes and through the valleys between mountains, rivers flow into the Atlantic and Pacific oceans. Facing south and west, the Pacific slopes are steep, scored by deep cuts and crags. On the Atlantic side, the slopes are smoother, a series of gentle hills tumbling to the Caribbean Sea.

The Talamanca-La Amistad Cordillera is a mountain range that forms part of Central America's spine. It stretches through the southeastern section of Costa Rica and on into Panama.

The World Heritage site that bears its name does not encompass just one or two protected areas, as is usually the case. Instead, it includes: two national parks—Chirripó, with 195 square miles (50,150 hectares), and La Amistad, with 540 square miles (139,929 hectares); two biological reserves—Hitoy-Cerere, with 35 square miles (9,044 hectares), and Barbilla, with a little under 40 square miles (10,000 hectares); and eight Indian reserves, including many archaeological ruins. In all, the Talamanca-La Amistad Cordillera covers about 3,860 square miles (one million hectares) under an assortment of programs. This makes it one of the largest protected forests in Central America.

Exuberant Vegetation
The tropical rain forest is the plant kingdom at its most exuberant. In temperate climates, 1 hectare (about 2.5 acres) of forest usually does not have more than ten tree species. More than one hundred kinds of trees grow on a hectare of land in the rain forest, and in some places the number is as high as two hundred. Hundreds of plants have not yet even been identified, and the importance of most of them is still unknown. The photos on the opposite page show three typical plants of the Panamanian tropics: a lignum vitae tree in flower, a white water lily, and the heliconia.

Degradation of the Environment
This sad picture (left) shows a tropical rain forest transformed by human activity. The soil is fragile and the plant cover on the ground is sparse. When the trees have been cut down and the earth plowed for the first and only time, the land must be abandoned. It becomes almost sterile.

23

A View of Two Oceans

The highest point in Talamanca-La Amistad is Chirripó Grande, 12,530 feet (3,819 meters) above sea level. Chirripó is also the highest point in Costa Rica. Different climates prevail at different altitudes in the cordillera. The climate, plus the area's rich volcanic soil, make the vegetation very diverse.

On the Caribbean (eastern) slope, the lower regions are swept with frequent mists blowing in off the sea. Up to an elevation of about 2,950 feet (900 meters), the land is covered with a dense, steamy jungle. The atmosphere is tropical, with plenty of palms, tree-sized ferns, and lianas (climbing plants) that reach the thickness of an adult's body. Above 2,950 feet, a new forest tapestry takes over—a majestic landscape of milk mushrooms, sweet cedars, and catkin.

On the lower regions of the Pacific (western) slope, there is a tropical deciduous forest. It is replaced here and there by grassy meadows.

At higher altitudes, woodlands of evergreens abound, composed of pines and Central American junipers. The forest is dotted with ferns, orchids, and begonias—plants that thrive in humid conditions.

The animal life is very diverse. Over 260 species of amphibians and reptiles and more than 400 kinds of birds have been counted. Among them are the quetzal (also known as the phoenix of the jungle), the crested eagle, the red-tailed hawk, the volcano hummingbird, and the black peacock.

Tropical Timeline

- **1513** Vasco Núñez de Balboa first sights the Pacific Ocean.
- **1538** Gonzalo Jimenez de Quesada, seeking El Dorado, conquers the Chibcha Indians of the Andes and founds the city of Bogota, Colombia.
- **1914** The Panama Canal opens, connecting the Atlantic and Pacific oceans.
- **1930** Construction of the Pan American Highway begins.
- **1980** Río Plátano National Park (Honduras) is declared a biosphere reserve. Panama names Darién a national park.
- **1981** UNESCO includes Darién on its list of World Heritage sites.
- **1982** The Río Plátano Biosphere Reserve is declared a cultural and wildlife reserve of the World Heritage. Chirripó and La Amistad national parks (Costa Rica) join the list of biosphere reserves.
- **1983** Chirripó and La Amistad national parks, as a cluster of biosphere reserves, are added to UNESCO's World Heritage list.

Talamanca-La Amistad
The Talamanca-La Amistad Cordillera is made up of national parks, biological reserves, and a special zone for Indian people. Its total area covers about 3,860 square miles (1,000,000 hectares). Its varied altitudes and climates, its strategic location, and the mixing together of plants and animals of North and South America have created a uniquely rich flora and fauna.

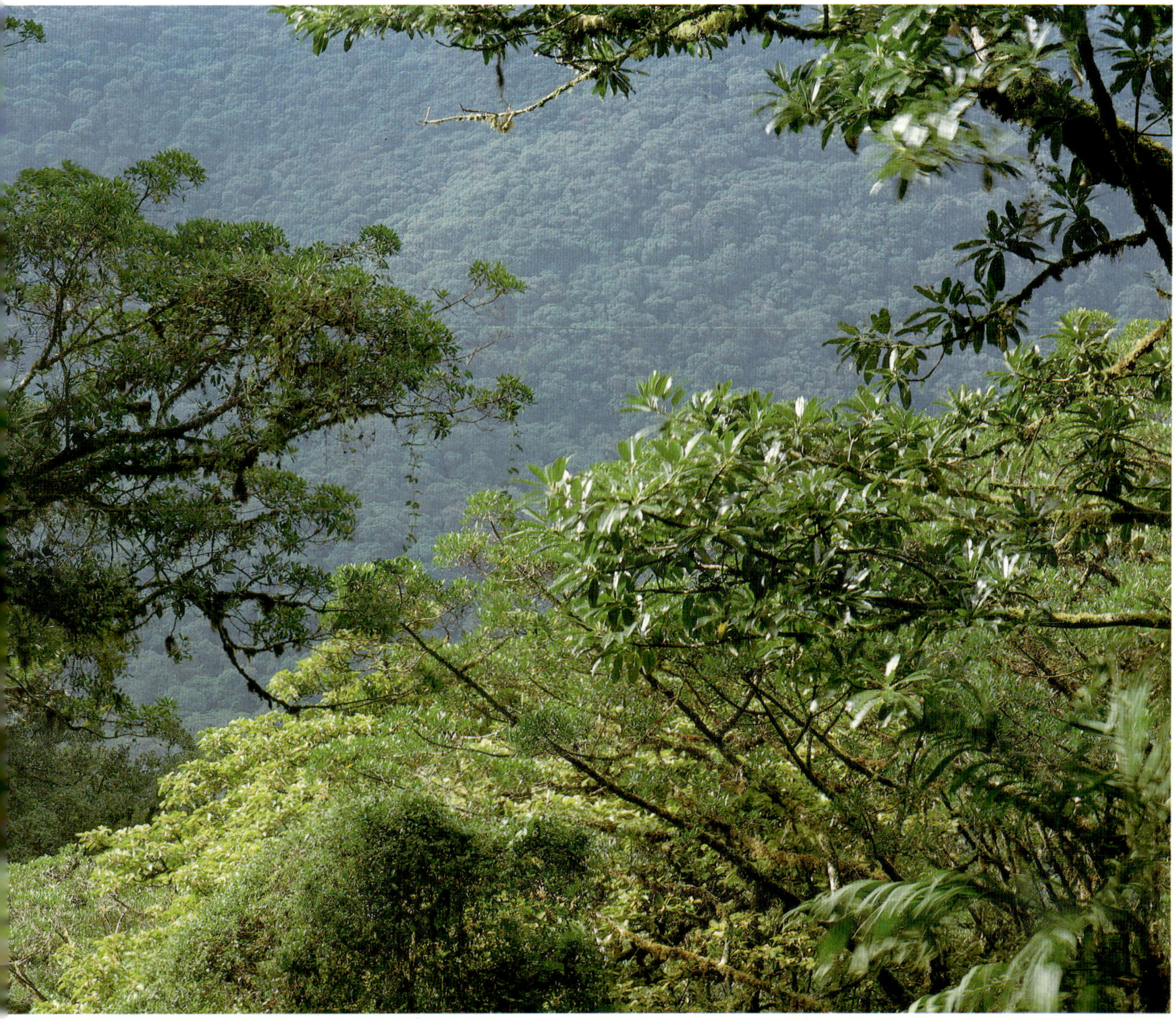

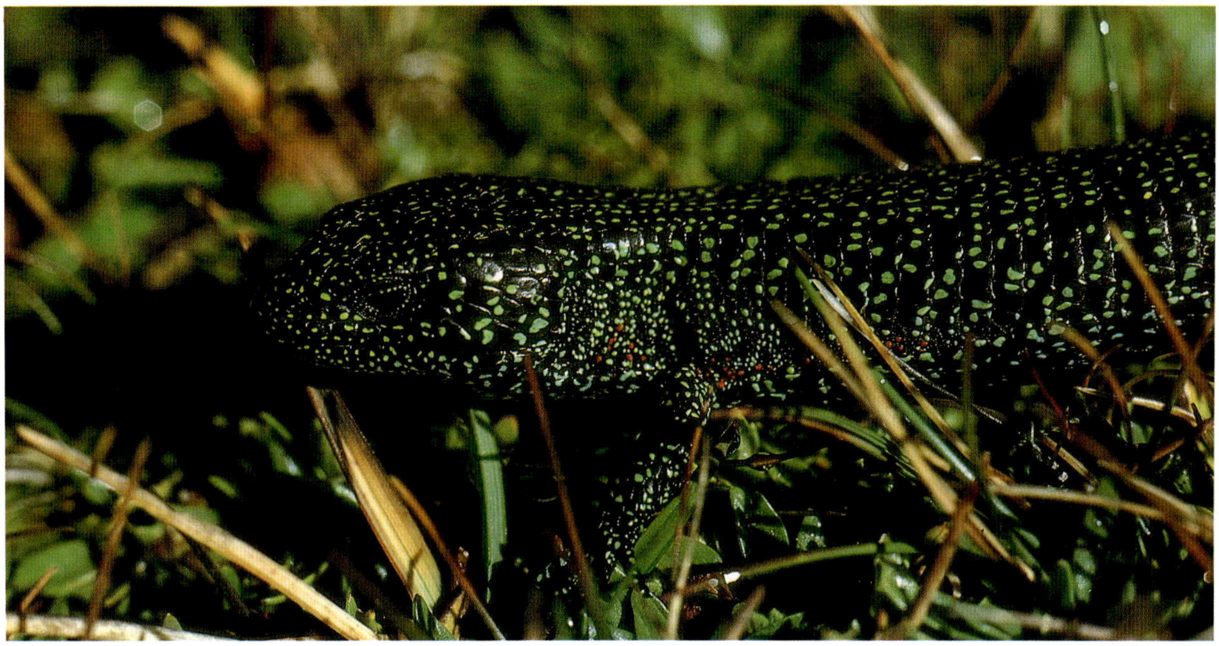

Some of the most stunning landscapes in the Talamanca-La Amistad Cordillera are its high plateaus. Above 9,840 feet (3,000 meters), there are meadows. The land is dotted with marshes and bogs created by the harsh climate and by the poor drainage of the soil.

The treasures of the heights are not limited to these plateaus. In Chirripó National Park are various landforms that are unique in all of Central America: glacial lakes, U-shaped valleys, terminal moraines, and glacial cirques.

These landforms were created by the movement of glaciers 25,000 years ago. These remains show us how this land was shaped, like all of North America, under massive icecaps thousands of years ago.

Some Neotropical Mammals

The treetops of the cordillera are home to one of the strangest mammals on earth: the sloth. On first glance, it looks like a chubby little bear, covered with a thick coat of long, rough, brittle hair. The sloth actually belongs to the order of Edentata, a group of animals that have few teeth or no teeth at all. Other *edentates* include the armadillo and the aardvark.

Thanks to the work of an American naturalist, William Beebe, we know that the sloth devotes less than 10 percent of its time to feeding, about 15 percent to slowly moving about, and the rest…to resting or sleeping!

Much has been written in European literature about bats and their habit of feeding on human blood. The bats of the Old World, however, are strictly insect and fruit eaters.

The same cannot be said for their Neotropical relatives, however. These bats feed on fruits and insects as well, but some species also dine on the blood of other animals. They rarely bother humans. (When they do, the greatest danger is infection, not loss of blood.)

The tapir is a relic of the remote past. It originated in Europe and Asia and eventually spread throughout South Asia and into the Americas. Tapirs live in marshy jungles and along riverbanks, entering the water at the least sign of danger.

Today most of the world's tapirs are found in Burma, Thailand, Vietnam, Malaysia, and Sumatra, as well as Central and South America. The tapir's clumsy movements and its tendency to leave conspicuous trails give it away to hunters, who kill it for its meat.

Darién National Park
Forming a bridge between the two continents of the New World, Darién National Park in Panama includes an exceptional array of habitats: sandy beaches, rocky coasts, mangrove swamps, wetlands, and tropical jungles. Almost completely cut off from the outside for thousands of years, these 2,200 square miles (570,000 hectares) shelter numerous threatened or endangered animal species, such as the ocelot (*opposite page, top*). Below is the Cana region in Darién National Park.

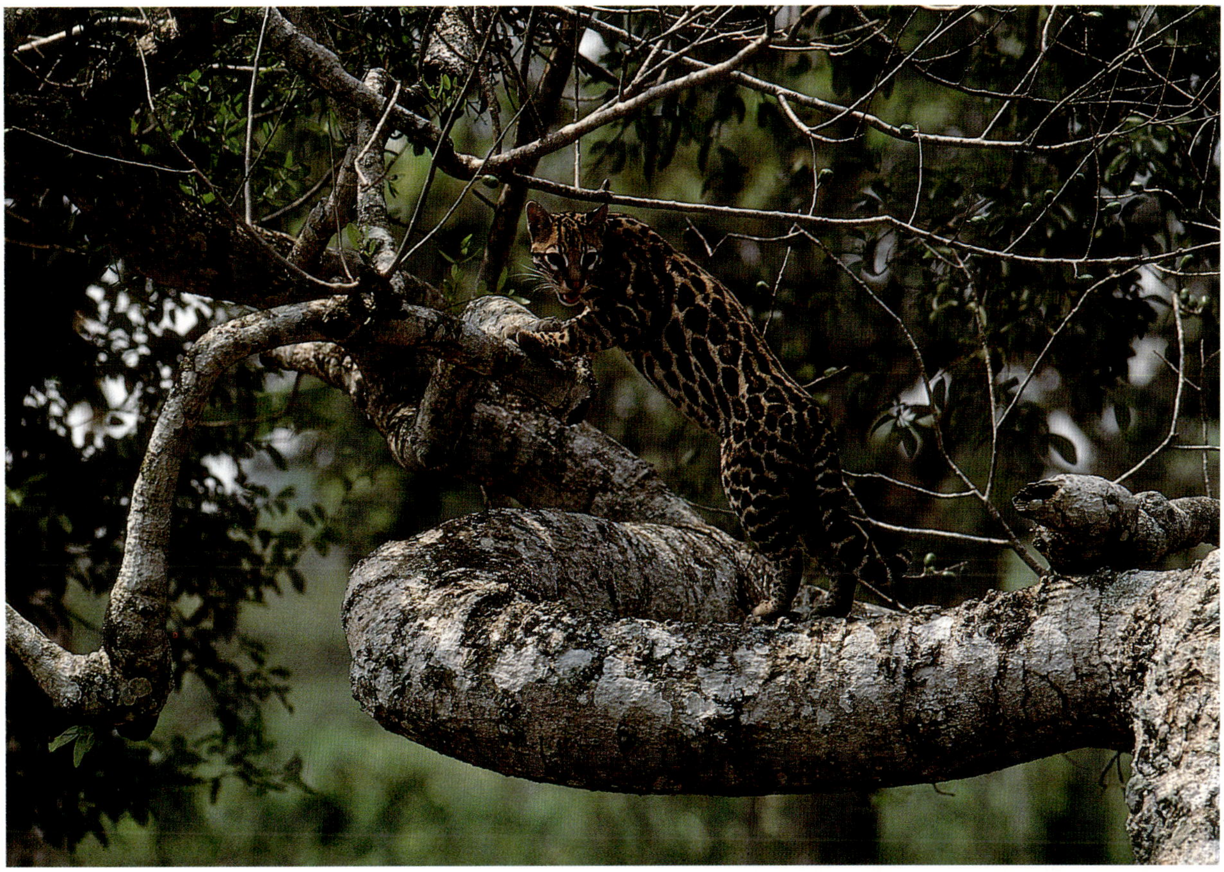

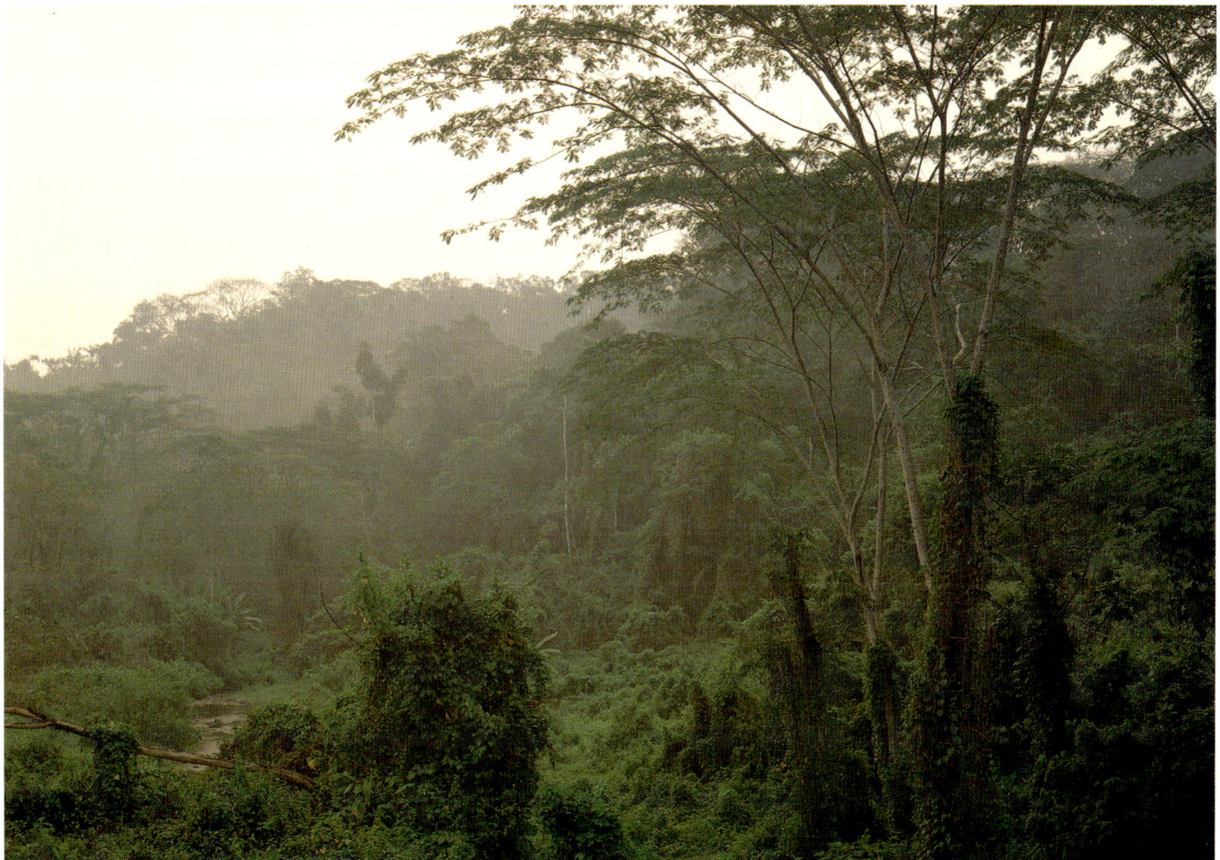

The Río Plátano Biosphere Reserve

Not often is a natural area completely isolated from population centers and major highways. Add unspoiled natural beauty and wonderfully diverse wildlife, and the area takes on exceptional character. If it also has a rare archaeological heritage, we have found one of the most remarkable places on the American continent.

Created in 1980 on 1,350 square miles (350,000 hectares) of land, the Río Plátano Biosphere Reserve is located in northeastern Honduras.

The reserve is flanked by the Paulaya and Sigre rivers and faces the Caribbean Sea and the Antilles Islands. Crowned by Mount Punta Piedra, 4,350 feet (1,326 meters) high, its slopes roll smoothly through wild terrain from the inland river basins to a lovely strip of coast.

The reserve is an endless swath of tropical forest, scarcely broken by rivers and plots of farmland around Indian settlements. Nearby along the coast lie two large bays or lagoons, with only tiny openings to the sea: Laguna de Ibans and Laguna de Brus. They are two of the largest bodies of water in northeastern Honduras.

Around the edge of the reserve is a buffer zone of 580 square miles (150,000 hectares). This region is intended to lessen the harmful effects of human activity outside the protected area.

The river system is the only transportation possible through the reserve. Vehicles cannot get through it. This also helps guard Río Plátano from invasion.

The Mangrove Barrier

The main ecosystem in Río Plátano is the subtropical rain forest. It receives 98 inches (250 centimeters) of rainfall every year and has a mean temperature of 79 degrees Fahrenheit (26 degrees Celsius).

Within the Río Plátano reserve are many different types of ecosystems. There are wetlands, a coastal plain covered with grasses and Caribbean pines, and an imposing forest along the rivers.

Finally, along the shore and the edges of the coastal lagoons, is a barrier of mangrove swamps. These swamps shelter a fascinating array of animals and plants.

The Tropical Rain Forest Put to Use

Every time we drink coffee, eat chocolate, or use something made of rubber, we are using products of tropical rain forests. If all the rain forests of Central and South America disappeared, all the wild varieties of cacao, rubber, and avocado would also vanish, along with many other species. The forests offer incredible benefits for the cure of diseases, for the climate of the entire earth, and for the genetic diversity of species. (*Opposite page, top*) The Plátano River in the Río Plátano Biosphere Reserve. (*Bottom*) One of the mountain streams in La Amistad National Park.

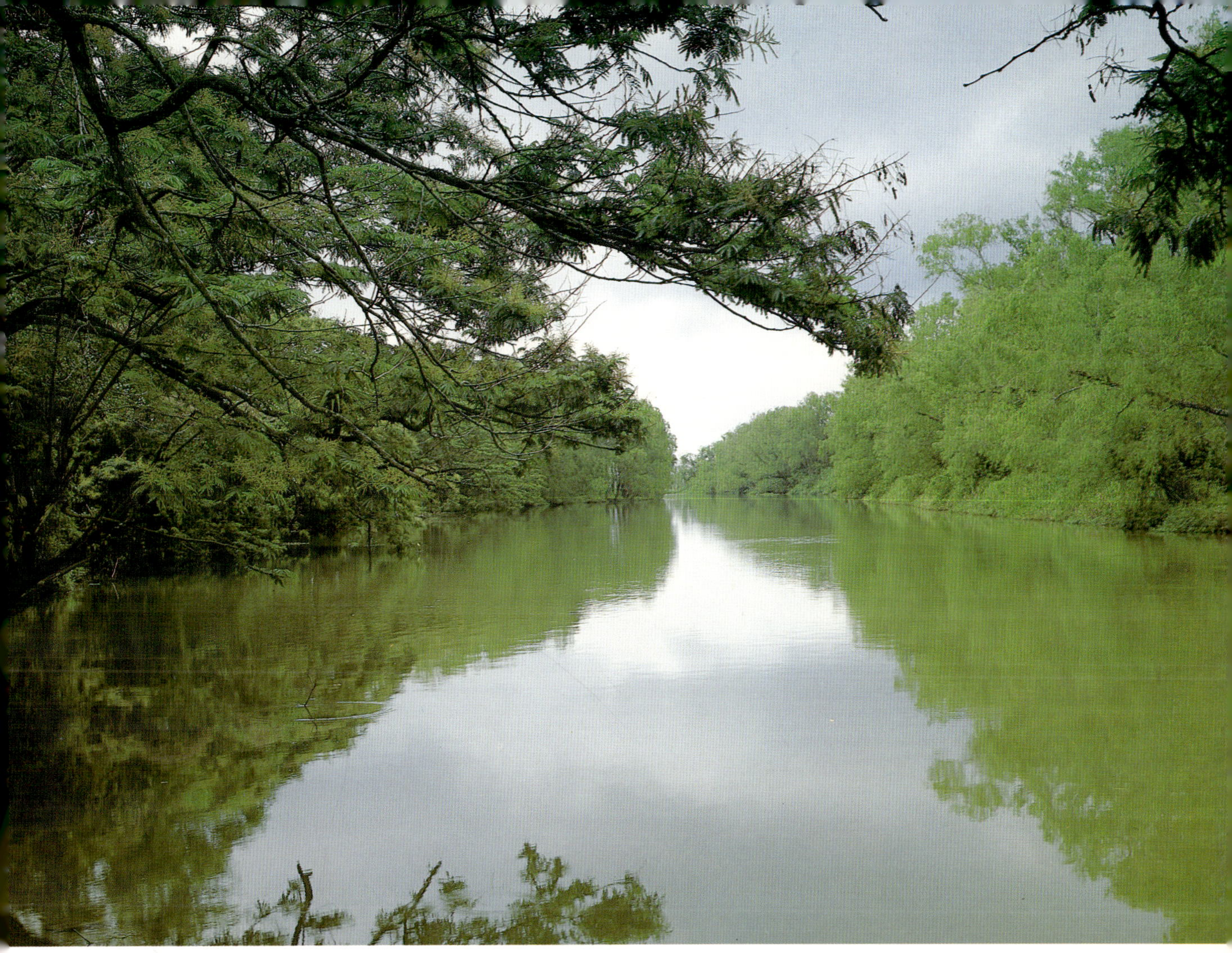

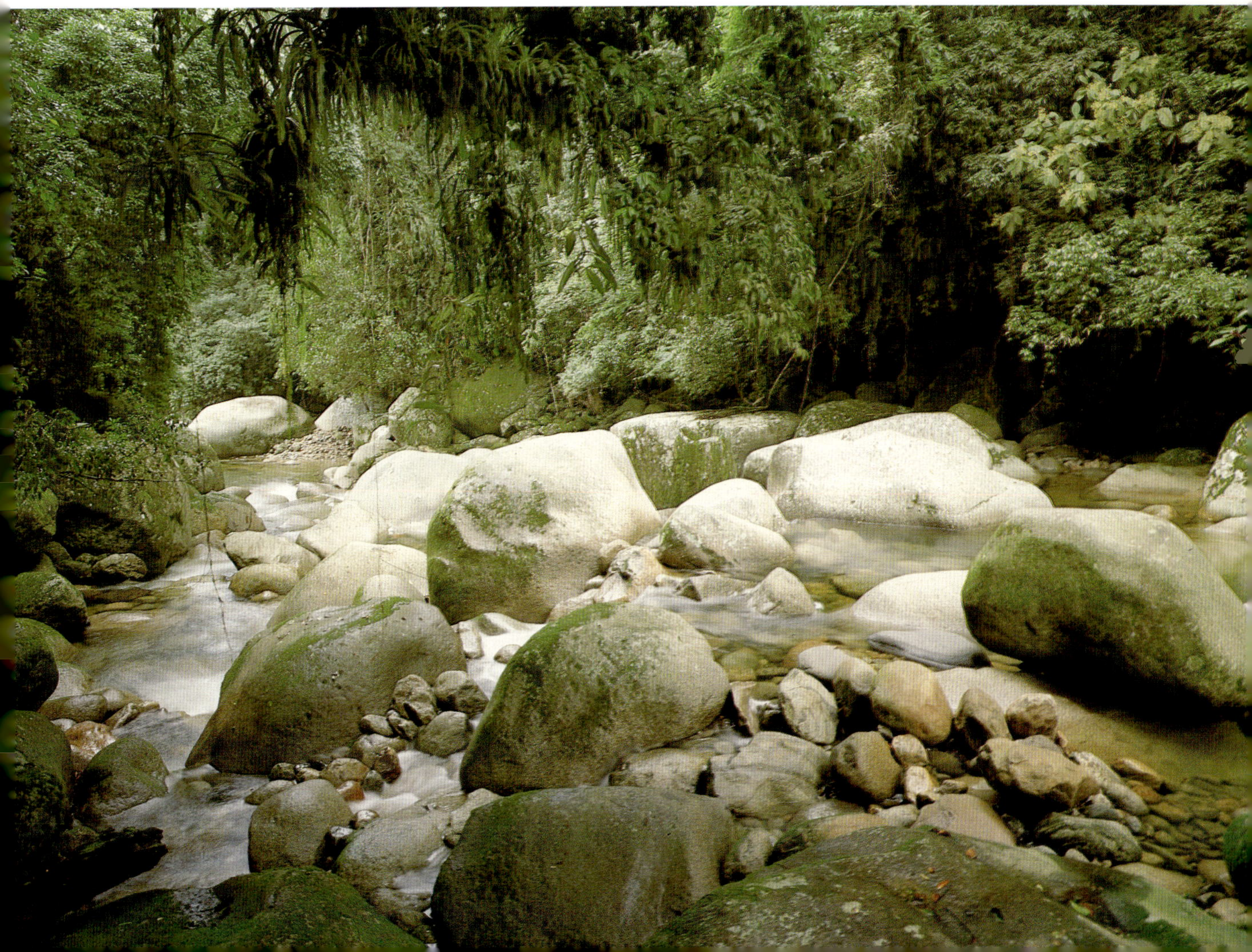

On approaching the reserve, the first thing the visitor notices is a narrow band of shore where it is almost impossible to tell where water ends and land begins. This is the mangrove swamp. It extends inland along deep channels and nearly clogs the entrances to the coastal lagoons. Swept by the daily action of the tides, the area teems with life. Its deep, salty waters attract many species. One that stands out is the red mangrove, a tree with thick external roots sprouting out from its trunk. Another is the coconut palm, noted for its fine, moisture-resistant wood. The mangrove swamp holds in place the soil and organic material that washes from the mainland. This causes the coastline to advance slowly into the ocean. Without the swamps, the sea would gradually wear the land away.

Among the animals of the mangrove swamp is the manatee. Manatees are large aquatic mammals. They are hairless and cigar-shaped, with front limbs modified to flippers. Manatees can weigh up to 1,400 pounds (650 kilograms), although they average about 450 pounds (200 kilograms). Their average length is 6.5 to 13 feet (2 to 4 meters). The manatee is strictly a vegetarian. Thanks to its streamlined form and rudder-shaped tail, it is a strong swimmer. On the other hand, if it runs aground on a beach, the manatee is defenseless.

Two other creatures of the mangrove swamps are the great American crocodile and the spectacled caiman, a smaller creature that looks something like a slender crocodile. Both have been hunted for their hides, which are prized for making shoes, purses, and other fine leather goods.

The Destruction of the Tropical Rain Forests

Honduras is losing over 300 square miles (80,000 hectares) of rain forest every year. Its ecosystem is degenerating faster than any other on earth. If this pattern continues, within 25 years its jungles will be nothing but a memory.

This makes it even more urgent that the Río Plátano Biosphere Reserve be preserved in its primitive state. It is the region's most important forested area. Nevertheless, specialists from the International Union for the Conservation of Nature (IUCN) have listed it among protected areas that are gravely endangered. Indian villages there have many social problems—particularly alcoholism—resulting from contact with Western civilization. These problems lead to the abuse of the rain forest—hence the IUCN's sad conclusion.

Tropical Rain Forest Alert!
The entire international community of conservationists—led by the World Wildlife Fund (WWF) and the International Union for the Conservation of Nature (IUCN)—is in a state of alert. They are warning the world that we must stop the destruction of rain forests throughout the earth. According to calculations, the slashing and burning of tropical rain forests destroys 42,500 square miles (110,000 square kilometers) per year. At this rate, they will have disappeared completely within a century. Conservationists urge the total protection of these exuberant jungles that have offered us so many gifts and still have so much to offer us in the years ahead.

These Sites Are Part of the World Heritage

Darién National Park (Panama): Its 2,200 square miles (570,000 hectares) are a vital stepping stone for the animals and plants that migrate between North and South America. It contains an exceptional variety of habitats, ranging from humid rain forests to coastal mangrove swamps.

La Amistad National Park (Panama): Extending over 850 square miles (221,000 hectares), this park was accepted as a World Heritage site in December 1990.

Talamanca-La Amistad Cordillera (Costa Rica): Eight different ecosystems are distributed over the region's 3,860 square miles (1,000,000 hectares). Among them is the high plateau. This is the only place in Central America where glaciers have left visible traces. The park includes eight Indian reserves, home to nearly 20,000 people from four different indigenous groups.

Río Plátano Biosphere Reserve (Honduras): This bastion of Central American jungle covers over 1,350 square miles (350,000 hectares). Here live 2,000 Miskito Indians. Fascinating archaeological remains survive within the reserve, such as the inscriptions on the "Painted Stones" and the ruins of the White City.

Glossary

archaeology: the study of past cultures by digging up and analyzing their remains

buffer zone: a neutral strip separating two areas

carnivore: an animal that eats other animals

carrion bird: a bird that feeds on dead animals

catastrophe: a disaster

cirque: a steep basin in a rounded mountain

conspicuous: easily seen

continent: one of the seven great land masses on the earth

cordillera: a mountain range

deciduous: having broad leaves that fall off seasonally

edentate: an animal with no teeth or few teeth

extinct: no longer existing

harpoon: a spear used to hunt fish and other animals that live in the water

invertebrate: an animal that does not have a spinal column

isthmus: a narrow strip of land connecting two larger land masses

living fossil: an animal that has remained the same through many periods of the earth's history

machete: a large, heavy knife

marsupial mammal: a mammal whose young are underdeveloped at birth and are sheltered in the mother's pouch; kangaroos and opossums are examples

moraine: a mass of earth and stone that has been carried along by a glacier

omnivore: an animal that eats both plants and animals

placental mammal: a mammal whose females have an organ called a placenta, which permits the young to develop fully within the mother's uterus

rodent: a small mammal with two pointy-edged front teeth

saga: a long tale

silt: a deposit of soil carried by a body of water

subcontinent: a large land mass that is smaller than a continent

subtropical: bordering on tropical lands

vertebrate: an animal that has a spinal column

viscous: liquid but very thick

volcanic: coming from or related to a volcano

Index

Page numbers in boldface type indicate illustrations.

Africa, 8
Andes Mountains, 6, 18
animals, 4, 8-10, **8-9,** 12, **15,** 16, **20-21,** 24, 26, **27,** 30
Asia, 26
Atlantic Ocean, 6, 22
Aztec Indians, 18
Balboa, Vasco Núñez de, 12
Bayano River, 12, 16
Beebe, William, 26
Cabeza de Vaca, Álvar Núñez, 4
Cana region, **11,** 18, **27**
Caribbean region, 16, 22, 24, 28
Chibcha Indians, 18
Choco Indians, 4, **6,** 16-18, **17, 19,** 22, **31**
Chucunaque River, 12, 16
climate, 14
Colombia, 12, 18
Columbus, Christopher, 4, 12, 18
continental drift, theory of, 6
Cortés, Hernando, 4
Costa Rica, **5,** 8, 18, **20-21,** 22, 24, **24-25, 29,** 31, **31**
Cuna Indians, 4, **13,** 14, 22
Darién Gap, 12
Darién National Park, **5, 6, 11,** 12-16, **13, 17, 19, 20-21,** 22, **27,** 31, **31**
El Dorado, 18
Europe, 6, 26
geology of Central America, 6, 8
gold, 18
Honduras, **7, 19,** 28, **29,** 30, 31, **31**
Inter-American Highway, 12

International Union for the Conservation of Nature, 30
La Amistad National Park, **5,** 22, **29,** 31
Lake Guatavita, 18
living fossils, 10
Magellan, Ferdinand, 12
map, **10**
migrations, animal, 6, 8, 10
Miskito Indians, 4, **7**
Neotropical zone, 10, 26
North America, 6, 10, 12, 26
Orellana, Francisco de, 18
Pacific Ocean, 6, 12, 22, 24
Panama, **5, 6,** 8, **11,** 12, **13, 17, 19, 20-21,** 22, **23, 27, 29,** 31, **31**
Perensemico River, **13**
Pizarro, Francisco, 4
Pizarro, Gonzalo, 18
plate tectonics, theory of, 6, 8
Quesada, Jimenez de, 18
Raleigh, Sir Walter, 18
Río Plátano Biosphere Reserve, **7, 19,** 28-30, **29,** 31, **31**
Rocky Mountains, 6
San Blas, 16
South America, 6, 8, 10, 12, 18, 26
South Sea, 12
Talamanca-La Amistad Cordillera, **20-21,** 22, **24-25, 29,** 31, **31**
Tuira River, 12, 16, 18
World Heritage sites, 6, 10, 22-26, 35
World Wildlife Fund, 30

Titles in the World Heritage Series

The Land of the Pharaohs
The Chinese Empire
Ancient Greece
Prehistoric Rock Art
The Roman Empire
Mayan Civilization
Tropical Rain Forests of Central America
Inca Civilization
Prehistoric Stone Monuments
Romanesque Art and Architecture
Great Animal Refuges
Coral Reefs

Photo Credits

Front cover: J. A. Fernandez/Incafo; p. 3: J. M. Barrs/Incafo; p. 4: Javier Andrada & J. A. Fernandez/Incafo; p. 5: J. M. Barrs/Incafo; p. 6: Candy Lopesino & Juan Hidalgo/Incafo; pp. 7-9: J. A. Fernandez/Incafo; J. A. Fernandez & Javier Andrada/Incafo; p. 11: J. A. Fernandez/Incafo; J. A. Fernandez & Javier Andrada; p. 13: J. A. Fernandez/Incafo; J. A. Fernandez & Javier Andrada/Incafo; p. 15: Javier Andrada & J. A. Fernandez/Incafo; p. 17: J. A. Fernandez/Incafo; J. A. Fernandez & Javier Andrada/Incafo; p. 19: J. A. Fernandez/Incafo; C. Lopesino & J. Hidalgo/Incafo; p. 22: J. A. Fernandez/Incafo; p. 23: C. Lopesino & J. Hidalgo/Incafo; Javier Andrada & J. A. Fernandez/Incafo; J. A. Fernandez/Incafo; p. 25: J. M. Barrs/Incafo; J. A. Fernandez/Incafo; p. 27: Javier Andrada & J. A. Fernandez/Incafo; J. A. Fernandez/Incafo; p. 29: J. A. Fernandez/Incafo; J. M. Barrs/Incafo; p. 31: J. A. Fernandez/Incafo; back cover: C. Lopesino & J. Hidalgo/Incafo; J. A. Fernandez/Incafo.

Project Editor, Childrens Press: Ann Heinrichs
Original Text: Alberto Ruiz de Larramendi
Subject Consultant: Dr. Francisco Dallmeier
Translator: Deborah Kent
Design: Alberto Caffaratto
Cartography: Modesto Arregui
Drawings: Olga Perez Alonso
Phototypesetting: Publishers Typesetters, Inc.

UNESCO's World Heritage

The United Nations Educational, Scientific, and Cultural Organization (UNESCO) was founded in 1946. Its purpose is to contribute to world peace by promoting cooperation among nations through education, science, and culture. UNESCO believes that such cooperation leads to universal respect for justice, for the rule of law, and for the basic human rights of all people.

UNESCO's many activities include, for example, combatting illiteracy, developing water resources, educating people on the environment, and promoting human rights.

In 1972, UNESCO established its World Heritage Convention. With members from over 100 nations, this international body works to protect cultural and natural wonders throughout the world. These include significant monuments, archaeological sites, geological formations, and natural landscapes. Such treasures, the Convention believes, are part of a World Heritage that belongs to all people. Thus, their preservation is important to us all.

Specialists on the World Heritage Committee have targeted over 300 sites for preservation. Through technical and financial aid, the international community restores, protects, and preserves these sites for future generations.

Volumes in the *World Heritage* series feature spectacular color photographs of various World Heritage sites and explain their historical, cultural, and scientific importance.